Industry and the Environment

A Report by CoEnCo

(The Committee for Environmental Conservation)

LONDON 1977

ISBN 0 903158 11 6

FRONT COVER: Smoke pollution (1964) Scunthorpe, Lincs.
Photograph by Aerofilms Ltd.

INSIDE FRONT COVER: Sailing Club, Shepperton. After rehabilitation of the marginal land a worked-out wet sand and gravel pit can become a recreational centre.
Photograph by Sand & Gravel Association

Foreword

THE MEN on the Moon found only a dead and silent world. They returned to an Earth full of life, colour, movement and sound. But now this living world of ours is seriously threatened by three inter-related dangers.

There is an exploding world population with an increasing need for food, land and raw materials. There is great pollution of water, air and land. There is destruction of the natural habitats of the living things that make up the ecological balance of which we are all part.

Scientists are working to find the answers, but everyone – everywhere – can help. This book has been compiled from information supplied by leading conservation organizations in this country. Its aim is to outline in general terms, appropriate to each subject, how our impressive industrial effort can more effectively be harmonized with action to protect our environment.

CoEnCo is conscious and appreciative of much that has been done through many areas of industry to avoid or minimize harm to the environment. Some industrial concerns have established their own environmental divisions; this bears witness to their lively awareness of the common ground that exists between themselves and the community and the need to keep environmental issues in the forefront when plans are made and executed.

CoEnCo believes, however, that through a wide range of industry there still is a desire for a statement on the hazards and for information on where to go for expert advice on dealing with them.

It is hoped also that this book will provide information on industry's conservation problems for the many educational bodies at all levels who are working in this field. It is good that, in co-operation with the Council for Environmental Education, industry is helping to promote an educational programme by which universities, training colleges and schools will be associated with the teaching and practice of conservation.

More than this, the Committee hopes that this book will stimulate further co-operation between Government, industry and the voluntary environmental movement so as to reach a better understanding of the problems and their solutions. If one product of this book were the creation of suitable machinery to achieve this, CoEnCo would be a very willing partner.

Craigton.

Chairman of CoEnCo

Contents

CHAPTER ONE

Conservation of Wildlife

Guidance on what industry can do for wildlife

INTRODUCTION

1 A thin layer of the Earth's surface – soil, water and air – the biosphere, forms the life support system for all living things, from the smallest herb to the giant Redwood tree, from invisible bacteria to the largest whale.

2 Because of its geographical position Britain has a very varied climate and geology. These two factors have produced a remarkable range of habitats with a corresponding variety of plants and animals. The lives of all of them inter-relate in an intricate and ever-changing pattern. Most people instinctively accept the fundamental importance of such natural components as grass, trees and fish to the continuance of a healthy life on our planet: after all, plants are our primary source of food. (We even need the microscopic fungus – yeast – to produce the alcoholic content of our beer.)

3 For many people wildlife is important for its aesthetic and social values alone. Thousands spend their leisure time enjoying the beauty of our countryside and observing wildlife in its natural habitats. But, in addition to the amateur, there are several hundred skilled people who work full-time in government and voluntary nature conservation bodies to monitor and manage the natural environment; and they have a Biological Records Centre in which to store data. In fact our wildlife is probably among the best documented in the world.

4 In the 1960s we saw a striking example of the adverse effects of man's interference with the balance of nature. Persistent organochloride pesticide (used in the important job of reducing crop pests) built up in the natural food chains so that, for example, seed-eating birds accummulated huge doses of the pesticide. Birds of prey – such as peregrine and sparrowhawk – feeding on these contaminated birds acquired lethal doses very rapidly. Fortunately, voluntary bans in this country on the use of these pesticides reduced the risks to wildlife and human life just in time.

5 Many living things, and certainly all those in danger of extinction, are already protected by law; this list includes many species of birds' eggs, birds, seals, deer and badgers, as well as certain insects, reptiles and plants.

6 To meet the need for more food, better quality but fewer strains of many agricultural crops (for example the garden pea) are being grown. We are now dependent on a few strains potentially at risk of destruction by new diseases. A national seed bank is maintained as a store of genetic material for breeding fresh strains, but as an insurance for the future we must conserve wild species.

HOW INDUSTRY CAN HELP WILDLIFE

7 Many of the decisions taken by industry will in some way affect wildlife and some of these could be very beneficial. For instance, there may be a piece of woodland around a factory, or a site that cannot be developed for some time. At little or no expense, it might be possible to improve this land in a way which would add to the interest and well-being of the staff and benefit the company's image – not to mention the wildlife itself. Some of the following suggestions may be helpful:

(a) Does the business own or control land, for example: woodland, flooded pits, waste and building land, lakes and streams which could with skilled advice be improved for wildlife?

(b) Is there any marginal land around a factory or other building which could be planted with native trees and shrubs to encourage birds and butterflies, and also to add to staff amenities?

(c) Are there places where wild flowers, nettles and other weeds could be left to grow without detracting from the efficient appearance of the premises?

Studying a find in Bishops Wood, near Stourport-on-Severn. This wood, owned by the Central Electricity Generating Board, surrounds a national grid electricity substation and has been opened to Worcestershire and Herefordshire schools for natural history and environmental studies.
Photograph by Central Electricity Generating Board.

Scrub clearance by the Conservation Corps. If the scrub is left it will outgrow everything else, causing a variation in wild Flora.
Photograph by H. M. Frawley.

A small nature reserve such as this will provide a natural home for many interesting and attractive species.

(d) Is there a derelict pond or stream that could be cleared and so once more become a home for wildlife? The Conservation Corps would be delighted to do the work. They would put such a clearance job into the hands of their Area Expert. Their address is Zoological Gardens, Regent's Park, London, NW1 4RY (01-722 7112).

(e) Is there safe land which could be used as a study site by a local school, also providing a better understanding of the firm's work?

(f) Would the staff enjoy setting up a study group of their own, so providing a link with the neighbourhood and emphasizing the firm's concern with environmental matters?

(g) Could nest boxes be sited in suitable places within the grounds?

(h) Where a firm is involved in activities such as mineral extraction, could reclamation be done in a way which would encourage wildlife?

ADVICE ON PLANNING FOR CONSERVATION

8 When planning any of the above, or new ventures such as the expansion of a factory or the building of new works or offices where there is likely to be real impact on the natural environment, there are two levels of effect to be considered. First, where activities will have a minor impact: by careful planning, and at relatively low cost, the impact can probably be reduced to an acceptable level so that the wildlife will not be unduly disturbed. Second, where activities will have a major impact: here it is important to get prior advice from experts both in landscaping and in wildlife conservation management in order to find out what the problems will be. Opposition is likely from a wide range of groups if major disturbance to wildlife will be caused, and early consultation and consequent planning should make it possible to minimize both objections to the proposals and threats to wildlife. The nature conservation organizations are aware that most industrial and commercial enterprises do not knowingly or deliberately destroy wildlife. If it is possible to get together at an early stage inexpensive and mutually beneficial plans can often be prepared.

9 The first people to be approached should be the regional officers of the Government agency for wildlife conservation – the Nature Conservancy Council. The regional officers and their staff are professionals who will be able to give constructive help, especially if they are involved in the initial planning stage. Their addresses can be obtained from the NCC Headquarters at 19/20 Belgrave Square, London, SW1X 8PY (01-235 3241).

10 The County Nature Conservation Trusts which are private bodies and employ full-time staff will always assist. Their addresses can be obtained from their parent body, the Society for the Promotion of Nature Conservation, The Green, Nettleham, Lincoln (0522 32326). Assistance and advice can also be obtained from the Royal Society for the Protection of Birds, The Lodge, Sandy, Bedfordshire (0767 80551).

PRINCIPAL CONTRIBUTOR

Council for Nature,
Zoological Gardens,
Regent's Park,
London, NW1 4RY (Telephone: 01-722 7111).

All the voluntary organizations noted above enjoy some form of grant aid from the World Wildlife Fund (British National Appeal), which has also financed the production of this book. The address of the Fund is 29-31 Greville Street, London, EC1N 8AX (Telephone: 01-404 5691).

Industry and the Environment

A Report and Handbook by CoEnCo

Industry and the Environment

A Report and Handbook by CoEnCo £1.85

This is something new – an essential practical aid to help industry play its part in our increasingly environment-conscious society. This is the first handbook ever produced using the knowledge and experience of leading conservation organisations in this country.

The approach is positive and the emphasis is on practical action. While acknowledging industry's own contributions to the protection of the environment this book sets out to meet a need for information as yet unmet. The handbook is illustrated throughout.

It is also hoped that this book will provide information on industry's conservation problems for the many educational bodies associated with the Council for Environmental Education.

Contents

he Committee for Environmental Conservation (CoEnCo)
9-31 Greville Street, London EC1N 8AX. Telephone: 01-242 9647

ISBN 0 903158 11 6

Order Form

The Committee for Environmental Conservation
31 Greville Street London EC1N 8AX

ase send me . . . . copies of
lustry and the Environment.
nclose a cheque for. . . . . .payable to
EnCo. (£1.85 post free, £1.50 for orders
10 or more)

BLOCK CAPITALS PLEASE
Name

....

Address

....

....

....

B

CHAPTER TWO

Preservation of the Land

The philosophy of good land stewardship

THE COUNTRYSIDE

1 Conservation has been defined as 'a positive evolving concept of co-partnership between man and nature'. The countryside, as it is to-day, has been fashioned by man to suit his needs over the centuries. It will continue to change as his needs change. To-day, in many cases, it is not what must be done so much as a failure to understand the effect certain developments may have on the landscape, or the living conditions of people, which causes problems.

2 Industrialists have so many aspects to consider when planning new schemes that the amenity/conservation view may be considered too low down in order of importance or, worse still, not at all. It is vital to pay attention to the 'continuity of supply'. The conservation anxiety is that there is a trend towards gratifying present demands with little or no thought to how our grandchildren will manage, or what the long-term effects on the countryside will be, for example, of the removal of the hedgerows, the pollution of our streams or of large-scale mineral workings.

3 The pressures on the countryside are enormous. It has been estimated that by the year 2000 there will be 59 million people and 29 million vehicles in Britain. It is therefore essential that the best use should be made of every acre of land. The nation cannot afford irresponsible development and land management; it cannot afford a quarter of a million acres of dereliction, much of which is beyond reclamation; it cannot afford a five-fold rise in traffic noise levels. The trouble is that conservation of the environment cannot be measured in economic terms alone. How does one value a beautiful view or a Norman church? Economic advantage is easily measured. Social cost is too often thought of as just a figure of speech.

4 In striving to protect the environment the conservation groups neither desire nor intend to put too much pressure on industry, or to increase unnecessarily the cost to the community. The fact is that industry can only pare its margins by so much. The community must realize this and be prepared to share the cost of environmental improvement. But very frequently environmental pressures lead to changes in manufacturing processes and other procedures which can produce unexpected benefits. For example: the after-use of wet sand, and gravel pits which instead of lying derelict now bring in good money from leisure facilities, and provide bird sanctuaries, too.

DERELICT LAND

5 It has already been stated that with the present pressure on our dwindling resources of agricultural land it is essential that the best use should be made of every single acre. The country cannot afford to have land lying in a state of dereliction. The official definition of classified derelict land until 1973 was 'land which is so damaged by industrial or other development that it is incapable of beneficial re-use without treatment'. The distinction between classified derelict land and non-classified derelict land is important, since government grants for clearing dereliction are available only for classified derelict land.

6 At present, land in active use, irrespective of its condition, is excluded from the definition. Even, for example, a 100-acre mineral working on which extraction is taking place on the last five acres with the remainder worked out and unrestored, is regarded as being in active use – and is therefore outside the definition. Any land which is subject to planning conditions governing its restoration is also excluded. Even if the planning conditions remain unfulfilled or are unfulfillable the worked-out land is not classifiable as derelict.

7 By 1974 it became clear that the government statistics of derelict land were inaccurate and under-

A mining site at Roddymoor (Co. Durham) before reclamation.

The same site after reclamation. *Civic Trust Award 1971.*

estimated at 97,000 acres, and a revised survey system for England was introduced. (The last annual record for Wales was in 1969 – then 20,000 acres; there is no record for Scotland.)

8 The new arrangements for England will go a long way towards assessing the nature and scale of land waste in England. There remains unsurveyed, however, land which is in a derelict or near-derelict state through poor management, neglect or speculation. Even so, the better and more comprehensive information should make it possible for a more realistic provision of financial resources to be allocated to clearing up the backlog of inherited dereliction. The problem is two-fold: to bring back into beneficial use derelict land caused by past activities, and to ensure that no more occurs from present or future land use.

9 Most industrial concerns do not accept any legal or financial responsibility for past dereliction. They point out that they cannot be held responsible to society for the failure of earlier industrial operators to meet recently introduced standards. However, inherited dereliction does affect communities where industry is still a major factor.

10 With greater public awareness of environmental problems, concern for prevention and/or rehabilitation of derelict land will inevitably increase. This awareness will result in a viable after-use of exploited land with a distinct disinclination to accept the use of greenfield sites when land already spoilt might be used for industrial building, even if perhaps at somewhat greater cost.

11 Beneficial after-uses can be roughly categorized as a return to agricultural land; the development of facilities for outdoor recreation; the enhancement of amenity and conservation value land: the provision of land for housing, industry, roads, etc.

A JOINT ROLE FOR INDUSTRY AND CONSERVATION ORGANIZATIONS

12 Industry can do much to facilitate the passage of its plans by consulting both the local authorities and the major conservation bodies. It is important to do this before a company is committed to a course of action so that any essential changes in plans and designs can be absorbed without upsetting the timetable unduly. There are nearly two thousand societies at national, regional and local level which might raise objections, not to mention private individuals. Whilst no amount of discussion will clear all objections, these can often be substantially reduced if the major national organizations are consulted at an early stage.

13 It is here that the Committee for Environmental Conservation can be of service by putting industry in touch with the organizations whose comments would be most helpful in each case.

14 The conservation interest stretches from the environmental effects of the production of the raw material right through to the design and manufacture of an article, to its marketing and, in many cases, its final disposal after use.

PRINCIPAL CONTRIBUTOR

The Council for the Protection of Rural England,
4 Hobart Place,
London, SW1W OHY (Telephone: 01-235 9481).

CHAPTER THREE

Water Resources and Disposal

A summary of the sources of supply, the structure of the industry and the conservation of water after use

RESOURCES

1 The use of water in England and Wales for domestic, industrial and agricultural purposes from the public water supply was about 290 litres (64 gallons) per person per day in 1971 and it is expected that demand will rise to between 425 and 480 litres (93 and 105 gallons) per day by the year 2001.

2 The average rainfall in England and Wales is approximately 7,700 litres (1,700 gallons) per person per day, of which about half returns to the atmosphere by direct evaporation or by transpiration through plants. Thus there should be ample for all foreseeable needs; but most of the rain falls in the hilly country to the west and north, more or less remote from the districts of high demand, and the rainfall is not evenly spread throughout the year. This pattern of rainfall led to the development of a system of storage in upland reservoirs with pipelines to the districts of demand. But it is becoming increasingly difficult to find new reservoir sites which are technically suitable and do not provoke strong agricultural, political or amenity objections. There are also increasing objections to the abstraction of water from the heads of rivers, so denying them the water which would normally flow down them.

3 In present practice the trend is towards the use of the rivers as aqueducts, with upland reservoirs used for regulating the flow, so that there are fewer floods and no very low flows. Water is then abstracted lower down the rivers near the centres of demand. The water that is used for regulating the rivers may not come only from surface reservoirs; some may come from ground water storage and this may be recharged artificially in times of high rainfall. In some instances water may be taken from one catchment and transferred by aqueduct across country to regulate a river in a different catchment. Possible transfers being investigated include River Tyne to the Tees and Wear, and Severn to Thames. Such transfers may have important chemical and biological consequences which must be studied before any scheme is implemented. Another idea that is being studied in several areas is to allow the rivers to flow more or less unimpeded to their mouths and store the flow in one or more reservoirs in the estuary. Theoretically almost the full flow of the river can thus be used.

THE WATER ACT 1973

4 Under the Water Act 1973 the 29 river authorities, 157 water undertakings and 1,393 sewerage and sewage disposal authorities responsible for the water services of England and Wales were merged into 10 Water Authorities. These Authorities have inherited all the responsibilities, assets and powers of the bodies they have replaced. They are responsible for water supply and conservation, sewerage and sewage disposal, prevention of river pollution, fisheries, land drainage and the recreational uses of the waters. The Act also created the National Water Council and the Water Space Amenity Commission. A Central Water Planning Unit is funded by the Department of the Environment, and a semi-autonomous Water Research Centre is responsible for fundamental and applied research on all the problems of the use and disposal of water. Proposals for revising the present structure of the water industry are now the subject of public debate.

THE CONSERVATION OF USED WATER

5 After use water can be discharged to sewers, rivers, on to land or to the sea. Every discharge has to be registered with and approved by a Water Authority. The Authority can specify the level of purity the proposed discharge must meet before consent is given and,

A reservoir and treatment works in the Peak Planning Board area. *Photograph by Elsam Mann & Cooper.* *Civic Trust Award 1965*

in the case of discharge to sewers, fix charges which take into account the extent of purification the sewage treatment plant will be required to carry out. Under the Rivers (Prevention of Pollution) Act of 1961 the Authority can, after due notice, change the specification of purity required for any discharge. The Control of Pollution Act 1974 made provision for the tightening up of the control of pollution. One of the aims is to legislate for the situation arising from the dual function of the Water Authorities, which are now responsible for supplying water, treating sewage and determining the extent and level of pollution from effluents – as it were gamekeeper and poacher roles. Unfortunately, the implementation of much of this Act has been postponed for financial reasons.

6 When water is used it is changed in quality, almost always for the worse, and largely returned to rivers. By suitable treatment the used water can be purified and recycled. The deterioration that takes place in domestic usage can be dealt with satisfactorily in modern sewage plants. The Water Authorities have taken over some 5,000 municipal treatment works in England and Wales, serving over four-fifths of the population. Some of these works are old and inefficient and some are overloaded. Many of the sewers conveying domestic waste are very old and cracked or corroded. In many districts storm water is conveyed to the sewage treatment works and can cause untreated sewage to pass into the rivers. Finally, in many coastal towns there is no sewage treatment and untreated sewage can be washed ashore by tidal movement. Conservationists must be concerned that all this is put right first of all; it must be brought under control by the local sewage plants as soon as funds permit.

THE ROLE OF INDUSTRY

7 Industry plays an increasingly important role in the use of the nation's water resources and the maintenance of the amenity, recreation and consumer uses of, as well as the environment associated with, water and water space. Apart from the statutory obligation on industry to keep levels of contamination of water discharged from industrial premises below specified standards it is very much in the interests of industrial concerns to do all that is possible to reduce water consumption and contamination.

8 The position with discharges from industry varies widely. Each industry will of course be in touch with its own supplier. Standards achieved along the Thames are high, making possible considerable re-use of water in the lower reaches and providing amenities for large numbers of people. By comparison the position on the river Trent is bad. Additional supplies of pure water have to be transferred to the East Midlands at great cost to make good the resulting shortage. It is the intention that over a period of years major improvements will be effected in the purity of the rivers of this country. The technology to achieve this is available but an equitable sharing of the costs will need to be worked out.

9 A useful start could be made now because it has been estimated that so many industries could, by installing recycling equipment that could soon pay for itself through reduced water charges, contribute to a nation-wide saving of our 'water capital'. In this way – and even without spending anything extra – industry could contribute considerably to the cause of conservation, because any percentage less water needed between catchment and sea means that much less countryside given over to reservoirs and associated headworks, not to mention reduction in pollution made possible through greater resort to purification techniques. Moreover, well kept pure waters make possible their own maximum use for amenity and recreation, and they enhance the environment. So industries using water are urged to consider (if they have not already done so) whether they could make an investment in conservation that would not only be economically profitable but also provide wider benefits. Savings achieved in this way might, in a critical drought situation, make the difference between some industries being shut down or not.

10 The more water used and the more heavily it is contaminated in use, the less recycling that can be practised. With reduced recycling more storage and treatment of water will be needed to meet the ever-increasing demands. This will inevitably raise water charges and, possibly, taxation. In this connection attention is invited to the Angling Foundation's scheme of Gold Medal Awards to Industry designed to encourage companies to install water-recycling and effluent-improvement systems. Information may be obtained from the Foundation's office at 145 Oxford Street, London W1R 1TB.

11 Industrialists are advised to inform themselves regularly and thoroughly of the position with regard to water consumption and effluent discharge wherever this arises in the industries in which they are employed or which they direct. As with so many other usages it is possible that wasteful practices may be brought to light. The nearest Water Authority will be ready with advice. Many cases of gross violation of the regulations arise from lack of attention and not by deliberate act or neglect.

12 Further information may be obtained from *Who's Who in the Water Industry*, published by the National Water Council, 1 Queen Anne's Gate, London SW1H 9BT. This is a directory of the Council itself, the Water Space Amenity Commission, the Water Authorities, the Water Companies, the Central Water Planning Unit and of many other official and public bodies.

PRINCIPAL CONTRIBUTOR

CoEnCo Water Sub-Committee

CHAPTER FOUR

Air Pollution

A survey of the statutory framework of control

INTRODUCTION

1 Air pollution means the presence, in the air, of contaminants which are harmful to our health, to animals, to all forms of plant life and to property. There is indisputable evidence that quite short exposure to high concentrations of some pollutants can cause serious and sometimes irreversible damage. It is therefore very important to reduce the danger to our work force and our countryside to the minimum. Nearly every country has air-pollution laws; Britain is one of the leaders in this important field.

2 The principal legislation enforced by local authorities in Britain (excluding Northern Ireland) is contained in the Clean Air Acts 1956 and 1968, and the Control of Pollution Act 1974, but some of the 'nuisance' provisions of the Public Health Act 1936, extended and amended by the Public Health (Recurring Nuisances) Act 1969, are applicable in specified circumstances.

3 The control of air pollution in Britain is essentially a team effort involving local authorities, the Department of Industry (Warren Spring Laboratory and the National Survey) the Department of Employment (including the former Alkali Inspectorate in respect of emissions to atmosphere from the 'scheduled processes', and the former Factory Inspectorate in respect of dust and toxic fumes, etc., emitted inside the factory) and Police Authorities (in respect of road vehicle emissions).

Smoke pollution (1964) Scunthorpe, Lincs.
Photograph by Aerofilms Ltd.

SMOKE EMISSION

4 The following regulations are concerned with the emission of smoke:

(a) New plant must be capable of substantially smokeless operation.

(b) The proposed installation of new furnaces must be notified to the local authority.

(c) There are permitted periods for dark and black smoke emission from chimneys, and these standards apply throughout England and Wales.

(d) The Permitted Periods Regulations apply whether or not premises are subject to a Smoke Control Order but, additionally, if in a Smoke Control Area the occupier must ensure that smoke emission is prevented or, where plant or premises are conditionally exempted, that smoke emission is minimized.

(e) A grant of seven-tenths of reasonable expenditure is normally available for the adaptation or replacement of domestic fuel burning and similar open fires in a Smoke Control Area.

(f) Dark smoke emissions from bonfires, etc., on industrial or trade premises are prohibited, subject to exemptions.

(g) The authorized officer must notify the occupier of premises as soon as possible of the existence of an offence, and must confirm his notification in writing within 48 hours.

(h) Smoke, other than smoke emitted from a dwelling, dark smoke from a chimney or dark smoke from industrial or trade burning, may be regarded as a statutory nuisance under the Public Health Acts.

CHIMNEY HEIGHTS

5*(a)* The occupier of premises must seek approval of a chimney height when:

(i) a new chimney is to be erected

(ii) an existing chimney is to be enlarged

(iii) a furnace is to be replaced by one of larger combustion capacity but only if the new or enlarged plant is to burn pulverized fuel; other solid matter at 100 lb. or more per hour; or liquid or gaseous matter at 1.25 million Btu or more per hour.

(b) An application for chimney height approval should preferably be on the form prescribed in the Clean Air (Heights of Chimneys) Regulations 1969 (Statutory Instrument 1969 No. 412).

(c) The local authority must give written decision within 28 days, and may prescribe conditions.

(d) The applicant can appeal to the Secretary of State for the Environment within 28 days of receiving the local authority decision.

(e) Chimney height approval is not required for certain exempted premises – mainly of temporary or experimental nature.

(f) All chimneys must comply with the relevant provisions of the Building Regulations 1972 as to the position of the terminal.

(g) In general the Departmental Memorandum on Chimney Heights is used, unless there are parameters more important than SO_2 emission which must be considered.

(h) In some circumstances, the developer of a tall building may be required to raise the height of the chimneys of an adjoining building.

THE CONTROL OF GRIT, DUST AND FUME EMISSIONS

6 Although 'grit' is defined in the Clean Air (Emission of Grit and Dust from Furnaces) Regulations 1971 as 'particles exceeding 76 microns in diameter', and 'fume' is defined in the 1968 Act as 'solid particulate matter smaller than dust', there is no definition of 'dust'. The requirement for grit arrestors to be installed in connection with new plant relates to those burning pulverized fuel, or other solid matter at the rate of one ton or more per hour.

HEALTH AND SAFETY: 1974 ACT

7 The main purpose of this Act is to provide a comprehensive and integrated legal and administrative system for securing the health, safety and welfare of people at work, and for protecting others against risks to health or safety arising from this work. To this end all the separate Inspectorates (including the Alkali Inspectorate) and agencies at present involved in these matters have been brought together in a Health and Safety Executive under a Health and Safety Commission which operates under the aegis of the Depart-

ment of Employment. (In its Fifth Report *(Air Pollution Control: An Integrated Approach)* the Royal Commission on Environmental Pollution has recommended, *inter alia*, the return of the Alkali Inspectorate to the direct control of the Department of the Environment.)

8 The Commission was set up on 1 October 1974 and the Executive on 1st January 1975. Initially these bodies will be responsible for enforcing the existing mass of legislation (including the Alkali Act 1906 but not the Clean Air Acts).

INFORMATION TO LOCAL AUTHORITIES

9 In accordance with Section 79 of Part IV of the Control of Pollution Act 1974, local authorities may obtain information about atmospheric pollution and about emissions from the smallest trade premises as well as the largest industrial complexes. The local authority may issue a notice to the occupier of premises requiring specific information which must be provided within the time allotted. This will not be less than six weeks. Alternatively, the local authority may ask the occupier to record data on its behalf or, with the occupier's consent, obtain the data itself.

10 The Act does not require the occupier of premises to take action before receiving a request from the local authority.

ROAD TRAFFIC POLLUTION LAW

11 Controls over emissions of gaseous pollutants from petrol-driven vehicles are implemented by the Motor Vehicles (Construction and Use) Regulations 1973. (For the purpose of limiting or reducing air pollution, the Control of Pollution Act 1974 enables the Secretary of State to issue Regulations imposing requirements about the composition and contents of any fuel used in motor vehicles.)

(a) Diesel Engined Vehicles

New diesel engines for use in road vehicles must be of a type certified to British Standard, BS AU 141a, or EEC Directive No 72/306, and engines must not be altered or adapted in a way which increases their smoke output.

(b) Petrol Engined Vehicles

Except for small three-wheeled cars and some exempted models, all petrol-engined cars and light commercial vehicles first used on or after 1973 and manufactured on or after 20 September 1973, must comply with the limits laid down for emission of hydrocarbons and carbon monoxide in Regulation No. 15 of the United Nations' Economic Commission for Europe.

Compliance with the Regulation must be attested by an approved mark, legibly and indelibly affixed to the vehicle. The attestation is carried out by the Vehicle Engineering and Inspection Division of the Department of the Environment.

PRINCIPAL CONTRIBUTOR

The National Society for Clean Air,
136 North Street,
Brighton BN1 1RG (Telephone: 0273 26313)

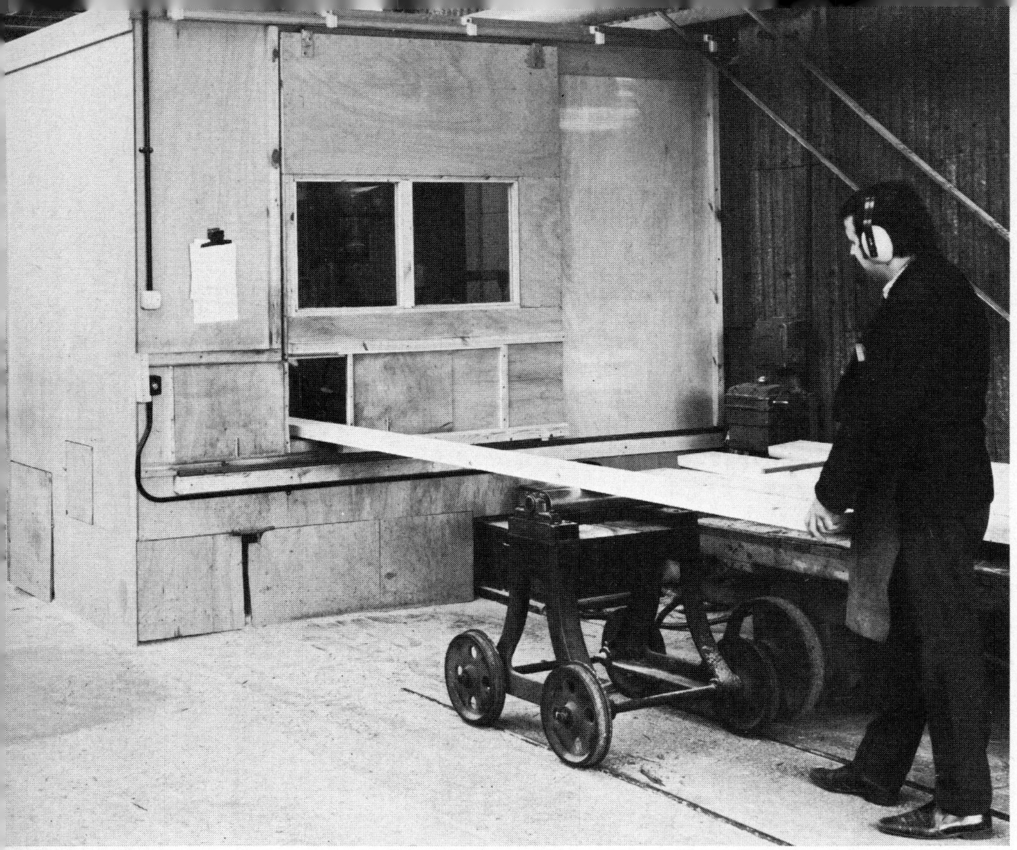

A hand resawing machine enclosure at the Cambridge mill of Cyril Ridgeon and Son Ltd. The sawyer is wearing ear protection (muffs). As a result of the enclosure noise levels are well below 90 dBA.
Photograph by Terence Reeve. Reproduced by courtesy of the Editor, 'Woodworking Industry'

CHAPTER FIVE

Noise

Industry's need to reduce it

INTRODUCTION

1 There is no longer any doubt that noise can be a severe health hazard to people at work. Habitual exposure to high levels of noise can cause serious loss of hearing. Not only is it well within the scope of the Health and Safety at Work Act 1974 but Part III of the Control of Pollution Act 1974 (which came into operation in 1976) was exclusively concerned with noise.

THE EFFECT OF LEGISLATION

2 The Health and Safety Act works through a system of Codes of Practice which have a legal position rather like that of the Highway Code: it is not an offence to break the code but it is *prima facie* evidence of an offence, and the onus is upon the defendant to prove that although he did not comply with the code he took

alternative steps which were equally effective. The statutory approval of Codes of Practice on Noise under the Act is only a matter of time; meanwhile officers of the Health and Safety Executive are concerned with getting all factory noise down below 90 dBA, and even lower where this is practicable. The Control of Pollution Act 1974 considerably tightened up the provisions in the Noise Abatement Act 1960.

3 For the first time industry is under pressure to reduce noise at work.

THE COST OF IMPROVEMENT

4 It is an important feature of the Health and Safety Act (as in the Control of Pollution Act) that the measures to be taken by industry should be on the basis of what is reasonable and practicable. One of the criteria for judging practicability is cost. Approximately 15 per cent of the workers covered by the old Factories Act are, on the latest estimate, exposed to noise levels of more than 90 dBA, 8 per cent of them continuously. Taking merely the latter group, an expenditure of £200 per man on noise reduction would cost industry in the region of £100 million. This cost may well come down if manufacturers improve their basic design – as a result of European Economic Community regulations and other factors.

5 Whatever the capital cost involved there undoubtedly will be some return for management. There is evidence that when noise levels are brought down, absenteeism is reduced and output improves.

6 But there is another cost to be reckoned – compensation to employees who bring successful actions in the High Court because of deafness caused by noise at work.

ACTION BY INDUSTRY

7 What must industry do to mitigate the enormous cost of noise? Clearly it must ensure that no further increase occurs. Staff must be trained in the protection of their hearing and in industrial noise control. Some Polytechnics and Universities conduct courses on Noise Measurement and Reduction. It is already known that there is an enormous variation in the susceptibility of people to hearing damage. Even at noise doses considered 'acceptable', a small percentage of people will eventually suffer severe loss of hearing. For this reason, hearing testing known as Audiometry is being introduced by many employers, both to ascertain the pre-employment hearing levels of new employees, and to monitor the progress of their hearing so that the extra-sensitive can be identified and taken off noisy work.

8 Wherever it can be achieved, the noise levels themselves must be brought down. Surveys of factory noise levels can be carried out by trained staff; experts can be called in to design noise-reducing modifications and equipment. The Noise Abatement Society always has available a panel of consultants to measure, advise, give evidence in Court, etc.

9 Only in intractable situations is a hearing protection device the answer and, in such cases, the actual ear defenders must be only a small part of the overall programme. The major part is to make employees aware of the effects of noise and the need to wear their ear defenders. In addition such measures as the identifying and marking of hazardous areas must be carried out. The only good ear defender is the one that is worn. Because of the mathematical nature of the noise dose unit, even lead and concrete ear muffs, if worn for only 90 per cent of the time can give no more than 10 dBA protection. Most ear muffs give about 30 dBA protection, but they must be worn for 99.99 per cent of time to achieve this.

10 Industrialists must now, as top priority, look to the designers of machines and buildings to help in solving the noise problem. The designers, for their part, should take acceptable noise level as an essential part of good design.

PRINCIPAL CONTRIBUTOR

The Noise Abatement Society,
6 Old Bond Street,
London, W1X 3TA (Telephone: 01-493 5877/9).

CHAPTER SIX

New Buildings and their Setting

The philosophy of the conservation of the built environment

1 Most people live in towns or cities – a built environment which can be squalid or glorious, oppressive or pleasant – and it is one of the paradoxes of our society that increasing wealth does not always lead to increasingly agreeable surroundings.

2 Our built environment, like our natural environment, is to-day under great pressure from the increase in population, from the tide of traffic and from redeveloping on an unprecedented scale. This pressure has generated a corresponding concern that the qualities which make for civilized surroundings are too frequently sacrificed in pursuit of other more obvious objectives. This concern is demonstrated by the recent rapid increase in local amenity societies dedicated to preserving and enhancing their localities, and also by the recent strengthening of legislation to control development.

3 All development affects the environment and, in an endeavour to ensure that private interests do not override the public interest, it is controlled by planning laws which are largely consolidated in the Town and Country Planning Act 1971. Applications for planning permission are made to local authorities who keep registers open to public inspection and anyone may make observations; for example, on whether a proposed development conforms with approved local plans, or on details of design. There is frequently public debate about proposed developments and important disputes are often referred to a public inquiry. It is therefore in the interest of the developer to know in advance that his proposals may be a matter of public interest.

4 Design is a creative act – it cannot be legislated for – it can only be encouraged. Education and example are therefore the principal tools of the voluntary bodies concerned with the built environment. Unless society as a whole values its surroundings we cannot expect buildings of quality. We get the surroundings we deserve.

5 Our heritage of buildings is protected in two main ways. Two hundred and forty thousand individual buildings are listed by the Department of the Environment as being of special architectural or historic interest, and these may not be altered or demolished without consent from the planning authority. But areas of character are often composed of buildings which have little or no individual merit but as a group their total effect *is* of special interest. These areas can be designated as Conservation Areas, and there are now some four thousand of them. Many are villages or the historic centres of towns. The intention is not that such areas should be frozen but that any new developments in the area should pay special regard to the existing character and should enhance it rather than diminish it. The concept is not one of mere preservation but of creative conservation – the aim being that, where necessary, existing buildings can be replaced but only if the replacement is an improvement. This requirement forces a consideration of whether the existing building cannot be adapted to suit a new purpose, a process which may require more thought and time to achieve but which may well be cheaper. The days of massive town centre redevelopment seem to be coming to an end – the public mood requires more sensitive treatment, grafting new buildings into old settings so that a sense of continuity is preserved.

6 Rehabilitation rather than redevelopment is also gaining acceptance for buildings of no great architectual distinction. Whereas some years ago it was common practice for local authorities to clear whole areas of run-down housing and build new estates it is now being realized that this is not necessarily cheaper and creates many social problems by disrupting existing communities. As materials become more expensive

The Abbeydale Works (Sheffield) before restoration (1969).
Photograph by Civic Trust.

The same Works after restoration.
Photograph by James Neill & Co. (Sheffield) Ltd. Civic Trust Award 1969.

A new factory for Nova (Jersey) Knit Ltd., in Monmouthshire (1971). *Civic Trust Award 1971.*

and the arguments for conserving them increase, rehabilitation rather than demolition is likely to be favoured even more.

7 It can therefore be seen that there is a coming together of three forces, aesthetic, social, and concern for resources, which now favour conservation of buildings.

8 One of the biggest threats to buildings and their settings has been the growth of traffic – cars because of their numbers and lorries because of their size and weight. Those concerned with the character of towns have therefore been drawn into consideration of transport policy and have advocated policies (which are now accepted by government) of restraining where possible car commuter traffic and encouraging the maximum sensible transfer of freight to rail.

PRINCIPAL CONTRIBUTOR

The Civic Trust,
17 Carlton House Terrace,
London, SW1Y 5AW (Telephone: 01-930 0914).

USEFUL ADDESSES

Ancient Monuments Society, 33 Ladbroke Square, London, W11 3NB
Georgian Group, 2 Chester Street, London, SW1X 7BB
Society for the Protection of Ancient Buildings,
55 Gt. Ormond Street, London, WC1N 3JA
Victorian Society, 1 Priory Gardens, Bedford Park, London W4 1TT

CHAPTER SEVEN

The Disposal and Reclamation of Waste

Review of recent studies and some recent statistics

PACKAGING

1 Attractive, effective packaging which is cheap to manufacture and easy to dispose of would serve both industry and conservation very well. Unfortunately it is rarely that simple: approximately 90 per cent of all packaging eventually becomes waste and its disposal and recycling present complex problems.

2 In 1972 Mr. Peter Walker, then Secretary of State for the Environment, sponsored a conference on 'Packaging and Litter'. Contributors included the conservation bodies, local authorities, organizations concerned with environmental problems, manufacturers and retailers of packaged goods, and designers and manufacturers of packaging and containers. Three working parties were set up to consider the impact on the environment of metal containers, glass containers, and plastics. The following paragraphs give a brief summary of their findings:

(*a*) *Metal Containers*

The British Tin Manufacturers' Federation submitted its report, *Metal Containers in the Environment*, in 1973. It was estimated that the U.K. production of these containers was £230 million in 1971, consuming 1 million tonnes of steel, 20,000 tonnes of aluminium and 8,500 tonnes of tin; that probably 95 per cent of all metal containers were disposed of by local authorities' refuse disposal services; that approximately 180,000 tonnes of used cans were recovered.

There would seem to be problems, both technical and economic, in expanding reclamation, and the working party felt that there was little evidence of overpackaging. It was, however, agreed that further research and development would be better directed to improving methods of collection, reclamation and recycling of metals in non-returnable containers rather than designing re-usable cans and setting up distribution and collection systems.

(*b*) *Glass Containers*

The Glass Manufacturers' report, *The Glass Container Industry and the Environmental Debate*, is a comprehensive account of the production, use and re-use of glass containers.

Glass contains the following materials: silica sand (50 per cent), soda ash (16 per cent), limestone (12 per cent), other materials (2 per cent) and cullet (20 per cent). Supplies of sand are said to be almost unlimited, environmental considerations apart, and 1.46 million tonnes were used in 1968. Three hundred and forty-five thousand tonnes of limestone are used annually apart from that used with salt to make soda ash, and both these minerals are virtually unlimited. Ninety-eight per cent of the constituent raw materials for manufacturing glass containers are indigenous. A total of approximately 2.5 million tonnes of glass was sold in the United Kingdom in 1971.

Glass makes up 9 per cent of refuse by weight. It is a material which, if uncontaminated, can be recycled without complication. But if the material is contaminated by 'foreign' cullet, i.e. crushed glass from another manufacturer, recycling is not so easy because of possible impurities. In 1971 430,000 tonnes of factory cullet (19 per cent) and 71,000 tonnes of 'foreign' cullet were used. It would be possible to reduce the problems of re-using container glass if all manufacturers were to rationalize the compositions they use. The new standards would then have to be approved by the British Standards Institution.

About 7,000 million glass bottles are produced in the United Kingdom each year. In 1972 returned bottles formed 17 per cent of all glass containers sold and non-returnable 10.4 per cent. The non-returnable bottle made its appearance in the mid 1960s, and its acceptance is said to be due to the increased ability to purchase goods and services, greater mobility and leisure and a general reluctance to return bottles. To this should be added the increase in self-service stores – 32,000 handling 71 per cent of retail sales – which do not want to be bothered with handling returnables, particularly as bottles no longer come in crates.

(c) Plastics

The Plastics Institute reported in January 1974. Its report on designing for disposability is entitled *Plastic Waste Disposal – Some Important Considerations.*

The industry's view is that plastic waste from packaging in municipal waste is approximately 2 per cent in weight now and may reach 6 per cent by 1980. It is not indestructible and does not cause pollution when disposed of by established methods. Plastics based on polyvinylchloride when burnt in municipal incinerators will not give rise to air pollution problems. The availability of degradable plastics (assuming all the technical and industrial problems can be solved) would lead to a reduction in the accumulation of plastic waste. Since the majority of plastics have oil as the source of their raw materials, the recycling and re-use potential of their wastes is particularly important. The industry believes that development work on recycling waste with the aid of salvage, pre-segregation or pyrolysis processes deserves greater attention; the nearer to the primary production the waste occurs the more it is likely to be recycled.

3 The British Plastics Federation has issued five booklets on facts about plastics in waste and waste disposal. These are entitled: *Degradability of Plastics; Plastics, their Contribution to Society and Considerations of their Disposal; Recycling, Re-use and Recovery of Plastics; The Storage and Collection of Refuse*, and *The Disposal of Solid Wastes.* These documents give an excellent insight into the plastics packaging industry and the problems arising from the disposal of plastic waste.

WASTE PAPER

4 Any reduction by industry in the use of paper is a positive contribution to conservation. Each tonne of paper uses 17 full-grown trees, 125 kg (275 lb) of sulphur, 162 kg (350 lb) of limestone, 266,900 litres (60,000 gallons) of water, 4 tonnes (9,000 lb) of steam and 255 kilowatt hours. Forty per cent recycled paper is now used in paper bags and low-quality wrapping paper. About 20 million paper bags are used daily. Sixty per cent in volume and 15 per cent in weight of all household refuse is paper and board; 2.1 million tonnes of waste paper accounted for 44 per cent of the total fibre used in the manufacture of new board and paper. Increasing world demand for timber makes it essential that even greater use should be made of waste paper.

5 The Department of Industry announced on 15 June 1976 a scheme providing £23 million for the paper and board and paper making manufacturing industries. A Waste Management Advisory Council Working Party is studying forecasts of supply and means of supply.

OTHER FORMS OF WASTE

6 Whilst packaging which makes up the bulk of refuse has been dealt with in some detail there are, of course, many other types of waste which could, and should, be reclaimed and recycled. The Government's green paper, Command 5727, *War on Waste – a Policy for Reclamation* gave a brief description of these in Chapter Five and includes ferrous metals, non-ferrous metals, rubber, textiles, chemicals, waste lubricating oil, mineral wastes and power station ash. Reference is also made to the British Research Establishment Paper CP 19/74 entitled *A Survey of the Locations, Disposal and Prospective Uses of the Major Industrial By-Products and Waste Materials.* Cmnd. 5727 gave the following figures:

(a) Ferrous Metals

In 1973 19 million tonnes of ferrous scrap were used. About half of this was recycled scrap from steelworks and foundries and the remainder was reclaimed material.

(b) Non-Ferrous Metals

In 1970 the cost to the nation of importing non-ferrous metals was over £700 million; the cost of copper alone was £270 million. In 1973 the United Kingdom consumption of the major non-ferrous metals was: aluminium 668,000 tonnes, copper 699,000 tonnes, lead 364,000 tonnes, zinc 391,000 tonnes, and platinum 18.7 tonnes, with scrap recovery rates being 26 per cent, 37 per cent, 62 per cent, 21 per cent and 58 per cent respectively. Recovery therefore plays an increasingly important part in the nation's balance of payments.

(c) Textiles

In 1973 136,500 tonnes of new wool were used and 16,000 tonnes recycled; 155,000 tonnes of new cotton and man-made fibres were used and 26,400 tonnes recycled. The industry is at present taking as much recycled material as it can absorb, and is experiencing difficulties with the increasing use of fibre blends.

(d) Mineral Wastes

The disposal of mineral wastes presents a special problem. In many cases the waste accumulates in unsightly spoil heaps leaving gaping holes which may be some distance away – making it expensive to fill the holes with the waste. This problem can only get worse until a use is found for this material, or until the cost of working aggregate reserves increases enough to make the use of waste an economic proposition. The magnitude of this problem can be judged from the following figures:

	TONNES PRODUCED ANNUALLY	TONNES STOCK PILED	TONNES USED ANNUALLY
Colliery Spoil	50m	3000m	7m to 8m
China Clay Waste	22m	280m	1m
Slate Waste	1.2m	300m	negligible

Some wastes, if they can be called that, are fully used. For example furnace clinker for concrete block making, blast furnace and steel-making slag for aggregates.

7 The aim should be for each industry to extract some beneficial use from its waste, bearing in mind that no commodity is limitless over and above a certain given rate of usage.

8 For more recent information on some aspects of the waste problem reference is invited to the First Report of the Waste Management Advisory Council and to the Council publications on waste paper collection by local authorities and on an economic case study of waste oil and its wider significance.

PRINCIPAL CONTRIBUTOR

The Council for the Protection of Rural England,
4 Hobart Place,
London, SW1W OHY (Telephone: 01-235 9481).

Fuel and heat and energy saving policies and devices can be a company's major contribution towards its own and the nation's prosperity. The Departments of Industry (01-215 7877) and of Energy (01-211 3000) are the Ministries principally responsible in this field.

A Rootes Goliath crane at British Rail Gosford Rail Depot, Coventry, unloads containers from the first full length train in Rootes factory-to-factory container service. (1970)
Photograph by Rootes.

CHAPTER EIGHT

Transport

A precis of some considerations

INTRODUCTION

1 Each industrial concern will endeavour to use methods of transport which are the cheapest, the most convenient and the quickest; but there are environmental factors which it is hoped that managers will not ignore when making their choice. At the end of this chapter there is a brief list of publications which, between them, highlight these environmental factors.

STANDARDS

2 In the case of the carriage of goods by road transport, managers will be acquainted with the Construction and Use Regulations which, in the interests of the community as a whole, impose certain standards – standards which are necessary not only in the interests of road safety but because of other aspects affecting our daily lives such as noise, the emission of pollutants and damage to buildings.

SPEED

3 Leaving on one side the humanitarian interests of road safety, driving vehicles too quickly is costly. First, there is the cost of fuel – a considerable burden on industry and the public generally. Second, there is the cost to the nation of accidents on the road attributable to over-speeding. According to the Government's recently published Consultation Document on Transport Policy road accidents cost the country about £850 million a year in police and hospital costs and lost production. This is a large slice of the Gross National Product. In so far as a significant number of accidents results from excessive speed those who control or use road transport will do well to keep the magnitude of the problem in mind – even if reduced to monetary terms alone. Stopping distances increase rapidly with speed (as paragraph 35 of the Highway Code makes clear) and much greater damage is done when an impact accident occurs involving heavier and faster lorries.

HEAVY LORRIES

4 Heavy Lorries are defined in the Heavy Commercial Vehicles (Controls and Regulations) Act 1973 (the 'Hugh Dykes Act'), as vehicles with an unladen weight exceeding three tons. It will be increasingly necessary for transport managers to take into account the effects of this Act and of Government proposals for a national system of routes for heavy lorries. Such lorries should be kept out of places where their presence cannot be tolerated on environmental grounds. Local authorities are under a duty to prepare regulations controlling the routing of heavy lorries within their areas. Local authorities may also make traffic regulations denying access to premises by such lorries for up to 24 hours a day on grounds of amenity. Such regulations will have to be taken into account by transport operators. According to the localities in which they operate, some commercial concerns may have to review their policies on the purchase of heavy lorries and to concentrate on the operation of lighter ones. Where heavy lorries are still in use it may become increasingly necessary to use district warehouses for the onward transmission of goods in smaller vehicles.

FUEL CONSUMPTION

5 In terms of fuel consumed per ton mile of freight carried, rail is more economic than road transport, and transport by water is even more economic than transport by rail.

RAIL TRANSPORT

6 Because of the need to-day to make the best possible use of fuel resources, it appears inevitable that the Government will have to place increasing emphasis on the use of railways for the transport of freight, and it could well be in the commercial interests of industrial firms to plan accordingly. There are powers under the Local Employment Act 1972 to assist in railway infrastructure on a regional basis and under the Railways Act 1974 to promote investment in private sidings.

PRINCIPAL CONTRIBUTOR

CoEnCo Transport Sub-Committee.

PUBLICATIONS FOR FURTHER INFORMATION

Transport and the Environment:
Committee for Environmental Conservation, 1973.

Lorries and the World We Live In: HMSO, 1973.

Transport – Co-ordination or Chaos:
Council for the Protection of Rural England, 1973.

Barges or Juggernauts – A National Commercial Waterways Development Projection:
The Inland Waterways Association, 1974.

Road and Rail Casualties in Great Britain in 1970:
G. C. Jenkins (printed in *Accident Analysis and Prevention*, Vol. 5, 1973).

Transport Policy: A Consultation Document,
Vols. 1 & 2: HMSO, 1976.

The Effects on Road Accident Rates of the Fuel Shortage of November 1973 and Consequent Legislation
Transport and Road Research Laboratory Supplementary Report 236.

CHAPTER NINE

Industrial Monuments

Aspects of the conservation of industrial machinery and buildings

ASSESSING THE PROBLEM

1 Unlike archaeological antiquities, industrial monuments may have only recently gone out of use or, in rare cases, they may still be in operation. Therefore the problem which confronts the owner of such a monument is usually not one of identification but of assessing its significance. In 1967 in an attempt to facilitate assessments the Government, in association with the Council for British Archaeology, established an Advisory Panel on Industrial Monuments. This Panel meets four times a year and has considered 1,166 sites since its inception. The majority of these sites have come to notice through the activities of the Industrial Monuments Survey which is co-ordinated by the Council for British Archaeology, with financial assistance from the Department of the Environment.

LEGAL PROTECTION

2 Industrial monuments share with conventional antiquities a common framework of legislative protection. Monuments which fall within the Ancient Monument Acts 1913-72 are normally recommended for scheduling to the Secretary of State by his advisers, the Ancient Monuments Board for England. Scotland and Wales have similar Boards to advise their respective Secretaries of State. Amongst those industrial monuments already scheduled are engine houses, furnaces, blowing houses, cranes, pottery kilns, glassworks cones, gasworks, pumping engines and colliery winding engines. However, scheduling is not the only form of protection accorded by the Department of the Environment. Its circular on *Industrial Archaeology* (AA 5193/45) stresses that:

> 'In addition to its activities under the Ancient Monuments Acts the Department has protected numerous industrial monuments by including them in the statutory lists of buildings of special architectural or historic interest compiled under the Town and Country Planning Acts 1947-71. These lists . . . protect the monument in that, once listed, it cannot be demolished or altered in a way that would affect its character, without the special consent of the local planning authority.'

AVENUES FOR PRESERVATION BY INDUSTRY

3 The Department of the Environment can, if it wishes, take monuments into the care of the Secretary of State – a procedure commonly known as guardianship. Amongst the industrial monuments already in guardianship are two eighteenth-century blast furnaces, one in Scotland and one in Wales, an engineering workshop attached to a slate quarry in North Wales, a lead mining site in mid-Wales, a nineteenth-century ironworks in South Wales, and a bobbin mill in Cumbria. The Department of the Environment also makes cash grants towards the preservation of outstanding monuments, and specialist advice is available from the professional officers of its Directorate of Ancient Monuments.

4 Since February 1973 the Science Museum in South Kensington, London, has administered a fund which can grant aid of up to £20,000 to local museums for the purchase, removal and re-erection of items of technological interest. Details of the operation of this fund can be obtained from the Science Museum itself.

5 In recent years local authorities have been playing a major role in industrial conservation. Some counties, such as Derbyshire, have appointed industrial archaeology officers and others have initiated or supported schemes to preserve such varied monuments as mine and canal pumping engines, lead crushing plant, industrial housing, and blast furnaces.

6 The National Trust is also active in the preservation of industrial monuments. It owns several water-

mills and windmills, half a dozen engine houses in Cornwall, some of which retain their engines, and an early textile mill and its associated model housing at Styal in Cheshire.

7 Private preservation trusts have provided a background and in many cases the foundation to much of this 'institutional' preservation. These bodies are usually charitable trusts set up to preserve specific monuments. With donations from private and industrial sources, they have preserved and restored steam-driven water pumping stations, windmills, watermills, tidemills, canal pumping engines, edge tool works, abandoned river ports, tilt hammer forges, textile mills complete with fulling stocks. A particularly notable example is the first sea-going iron ship, Brunel's *Great Eastern*.

8 A somewhat different kind of preservation trust is the open-air museum. Of those with an industrial emphasis some, such as Beamish in County Durham, are largely local authority financed, while others, such as the Ironbridge Gorge Museum, depend very much more on industrial contributions. These museum sites have the advantage of being able to accommodate bulky industrial material which, for various reasons, cannot be preserved *in situ*.

9 Industry too is playing a part in preserving its heritage. Only the management can identify a piece of equipment, perhaps still in operation, as of unique historic interest. British Rail and British Waterways have established national museums; the National Coal Board has a private museum at Lounds Hall in Nottinghamshire and British Gas is collaborating in the formation of two site museums. The British Steel Corporation has given valuable support to various preservation trusts. In the private sector the ceramic industry around Stoke-on-Trent has supported Gladstone Pottery Museum, housed in a typical nineteenth-century pottery; other companies have established their own museums such as the Museum of

Gladstone Pottery Museum, Longton, Stoke-on-Trent. Conservation of industrial premises with the support of private industry.

Glass at Pilkington's and English China Clay's open-air museum at Wheal Martyn.

NEW BUILDINGS AND RE-DEVELOPMENT

10 Factory expansion or a change in waste disposal plans may bring additional land into use. The industrialist will want to make sure that the changed use of the land will not destroy something of historic interest to the nation.

11 Archaeologists should be brought in at an early stage so that the area can be carefully checked before work commences. In the event of such a survey producing evidence of a major archaeological site it is not too late for plans and construction programmes to be modified, without added expense, to permit proper excavation and even preservation. The Roman fort at Dover with its unique painted room was preserved as a result of the sympathetic attitude of the contractors to the archaeologists who found it.

12 Archaeologists have developed rapid methods for the excavation and recording of such finds. The technique of 'rescue archaeology', properly applied and with goodwill on the part of the developer and of the archaeologist, can ensure a minimum of disruption and expense. It should not be overlooked, moreover, that insurance to cover such contingencies is now available to the developer.

13 In spite of close collaboration between industry and archaeologists cases will sometimes arise where although the importance of a site discovered is such that preservation in the public interest is felt to be desirable, this would be uneconomic for the industry concerned. In such cases it rests with industry to decide whether it has an obligation that transcends productivity and profitability. Notable precedents exist, such as the careful lifting and resiting of the Mithras temple from Roman London. The Department of the Environment is sympathetic to appeals for assistance when cases of this kind arise; nevertheless the main responsibility must inevitably rest with industry itself.

RESEARCH

14 The assessment of significance is not only undertaken by the agencies already mentioned. Professional institutions such as the Civil Engineers have panels to identify monuments of specific interest. Some industries are also trying to establish what is worth preserving.

15 In September 1973 the Association for Industrial Archaeology was formed. It promotes regional and specialist surveys and supports bodies involved in the preservation of industrial monuments. The Association has a wide membership and welcomes the interest of institutions and individual industrialists.

PRINCIPAL CONTRIBUTOR

The Council for British Archaeology,
7 Marylebone Road,
London, NW1 5HA Telephone: (01-486 1527).

USEFUL ADDRESSES

Directorate of Ancient Monuments,
Fortress House,
23 Savile Row,
London, W1X 2AA.

Hon. Secretary,
Association for Industrial Archaeology,
c/o Ironbridge Gorge Museum Trust Ltd.,
Southside, Church Hill,
Ironbridge,
Telford, Shropshire, TF8 7RE.

The Survey Officer,
Industrial Monuments Survey,
Council for British Archaeology,
School of Humanities and Social Sciences,
University of Bath,
Bath, BA2 7AY.

The Science Museum,
South Kensington,
London, SW7 2DD. (Telephone: 01-589 6371)

A textile mill renovated and now with the Industrial Museum, Bradford Metropolitan District Council.
Photograph by Bradford Metropolitan District Council. Civic Trust Heritage Year Award 1975.

APPENDIX

CoEnCo Committee Membership

Chairman The Rt. Hon. Lord Craigton, CBE

Vice-Chairman Mr. Stanley Cramp, OBE, FZS, Council for Nature

Treasurer Mr. D. J. Nuttman, World Wildlife Fund (BNA)

Anglers' Co-operative Association	Mr. Ken Sutton
Civic Trust	Mr. Nigel Haigh
Conservation Society	Mr. Edward Dawson
Council for British Archaeology	Mr. H. Cleere
Council for Nature	Miss Melinda Appleby Captain Sir Thomas Barlow, BT, DSC
Council for the Protection of Rural England	Mr. A. F. Holford-Walker, MA
Council for the Protection of Rural Wales	The Baroness White
Joint Committee: *Society for the Protection of Ancient Buildings* *Georgian Group* *Victorian Society* *Ancient Monuments Society*	Mr. M. Beaumont *ad interim*
National Society for Clean Air	Rear-Admiral P. G. Sharp, CB, DSC
National Trust	Mr. M. Beaumont
National Trust for Scotland	Mr. J. C. Stormonth-Darling, CBE, MC, WS
Noise Abatement Society	Mr. J. Connell
Ramblers' Association	Mr. A. Mattingly
Soil Association	Dr. E. F. Schumacher, CBE
Youth Hostels Association	Mr. J. Corlett, MSC, FIBiol
Ulster Society for the Protection of the Countryside	Mr. W. M. Capper

Observers

Commons, Open Spaces and Footpaths Preservation Society	Mr. Paul Clayden
Nature Conservancy Council	Mr. W. J. Midgley
Royal Society of Arts	Mr. Ian M. Hurrell, MVO
Town and Country Planning Association	Mr. David Hall
Pedestrians' Association for Road Safety	Mr. G. C. Jenkins, BA, FCIS

Secretary Mr. F. D. Webber, CMG, MC, TD

Asst. Secretary Mrs. Susan Dawson, BSC

Address 29-31 Greville St., London, EC1N 8AX (01-242 9647)
(Registered Charity No. 263154)

CoEnCo Objectives and Role

History and Aims of Constituent Bodies

THE COMMITTEE for Environmental Conservation was formed in 1969 with the following terms of reference:

To consider matters of national importance affecting the environment by providing a forum at which common problems, transcending the terms of reference of any one body represented, may be examined by representatives of the major national voluntary amenity and conservation organizations;

to promote concerted action by the voluntary bodies;

to distinguish and, where possible, settle points of conflict between the different environmental interests;

to identify areas of overlap or gaps in coverage;

to encourage the best use of existing resources;

to initiate such action and make such recommendations as may be necessary to further liaison between organizations at national, regional and county level.

To this end the Committee may appoint sub-committees or working parties and co-opt persons to these bodies as may seem desirable to provide the specialist information required.

It may set up such common services as may be deemed necessary, and it should in suitable cases encourage or promote the establishment of regional and/or county co-ordinating committees.

1 Joint Committee

Society for the Protection of Ancient Buildings
Founded by William Morris in 1877. Its interests cover ecclesiastical and secular buildings both large and small, including also mills, barns, dovecotes and bridges. The SPAB advises on all problems affecting old buildings; in particular, it offers technical advice on their treatment and repair. It publishes information on the history and care of old buildings, features and fittings and circulates a detailed index of threatened properties to prospective purchasers.
Address: 55 Great Ormond Street, London, WC1N 3JA

Ancient Monuments Society
Founded in 1924, the Society exists for the sole purpose of conserving our historic buildings and places and objects of historic interest, and mounts campaigns to increase the number of conservation areas which, through legislation initiated by the Society, are now protected.
Address: 33 Ladbroke Square, London, W11 3NB

Georgian Group
The Georgian Group was founded in 1937 to awaken and direct public opinion to the urgent need to protect the nation's steadily diminishing Georgian heritage. It has therefore fought for nearly forty years to save 18th century buildings of all kinds from demolition and is now one of the national amenity societies to which local planning authorities have been directed to refer all applications for demolition of listed buildings.
Address: 2 Chester Street, London, SW1X 7BB

Victorian Society
The Victorian Society was founded in 1958 to encourage the appreciation of Victorian art and architecture. Its aim is to preserve 19th and early 20th century buildings and industrial monuments. It is consulted on all listed building applications, proposes buildings to be listed and is currently engaged in a scheme to assist local authorities in the designation and enhancement of conservation areas.
Address: 1 Priory Gardens, Bedford Park, London, W4 1TT

2 National Society for Clean Air
Formed in 1889. The objects of the Society are to promote clean air in the U.K. by creating an informed public opinion on its importance and by other activities.
Address: 136 North Street, Brighton, Sussex, BN1 1RG

3 National Trust

The National Trust is a charity founded in 1895 and incorporated by an Act of Parliament in 1907. Its mandate in that Act is 'to promote the permanent preservation for the benefit of the nation of land and buildings of beauty or historic interest'. The same Act gave the Trust the power to declare its property 'inalienable'.

Address and title: The National Trust for Places of Historic Interest or Natural Beauty,
42 Queen Anne's Gate, London, SW1H 9AS

4 Ancient Monuments Society *(1924 – see under 1)*

5 Ramblers' Association

Founded 1925. Objects: protection of rights of public on country paths; preservation of the beauty of the countryside; the securing of the right of public access to the open country.

Address: 1-4 Crawford Mews, York Street, London, W1H 1PT

6 Council for the Protection of Rural England

Founded in 1926 to organize concerted action to secure the improvement, protection and preservation of the countryside and its towns and villages; to act as an information and advisory centre and to arouse and educate public opinion on matters affecting the visual amenities and usage of the countryside.

Address: 4 Hobart Place, London, SW1W 0HY

7 Council for the Protection of Rural Wales

Founded in 1928 with the same aims and objects and similar organizational structure as CPRE.

Address: 14 Broad Street, Welshpool, Powys, SY21 7SD

8 Youth Hostels Association (England and Wales)

Founded 1930. Object: to help all, especially young people of limited means, to a greater knowledge, love and care of the countryside particularly by providing hostels or other simple accommodation for them in their travels. Twenty-two of the Association's 260 hostels provide special facilities for field studies. There are Youth Hostels Associations in Scotland and Northern Ireland.

Address: National Office, 8 St. Stephen's Hill, St. Albans, Herts., AL1 2DY

9 National Trust for Scotland

A charity founded in 1931. To promote the permanent preservation, for the benefit of the nation, of places and buildings in Scotland of natural beauty or historic interest.

Address: 5 Charlotte Square, Edinburgh, EH2 4DU

10 Ulster Society for the Preservation of the Countryside

Founded 1937. To safeguard the amenities of Northern Ireland. To arouse and educate public opinion to this end. To act, either directly or through the members of affiliated bodies, as a centre for giving or obtaining advice and information on such matters.

Address: West Winds, Carney Hill, Holywood, Co. Down, BT18 0JR

11 Georgian Group *(1937 – see under 1)*

12 Council for British Archaeology

Formed in 1944 for two purposes. It provides liaison between archaeological societies, the universities and state organizations concerned with the preservation of antiquities, and assists the individual archaeologist, whether amateur or professional.

Address: 7 Marylebone Road, London, NW1 5HA

13 Soil Association

Founded 1946. To bring together all those working for a fuller understanding of the vital relationships between soil, plant, animals and man. To initiate, co-ordinate and assist research in this field. To collect and distribute the knowledge gained so as to create a body of informed public opinion.

Address: Walnut Tree Manor, Haughley, Stowmarket, Suffolk, IP14 3RS

14 Anglers' Co-operative Association

Founded in 1948 with the twin objectives of campaigning against water pollution and enabling, through the provision of funds, citizens to exercise their common law rights in the Courts by seeking injunctions to stop pollution and in claiming damages for financial loss and/or loss of enjoyment caused by pollution.

Address: Midland Bank Chambers, Westgate, Grantham, Lincs., NG 31 6LE

15 Civic Trust

Founded 1957. The aims of the Trust are to encourage high standards in architecture and planning and to preserve features of beauty or of historical interest; to stimulate public interest in these aims it encourages the formation of local amenity societies of which it maintains a register.

Address: 17 Carlton House Terrace, London, SW1Y 5AW

16 Council for Nature

Founded in 1958. The national representative body of

the voluntary natural history movement in the U.K., acting on behalf of naturalists and others to further the study and conservation of nature. The numerous bodies it represents include the Society for the Promotion of Nature Reserves (now Conservation), founded in 1912, which co-ordinates the activities of the County Naturalists' Trusts, the Royal Society for the Protection of Birds, founded in 1889 to encourage the better conservation and protection of wild birds and the Botanical Society of the British Isles, founded in 1832 to study the taxonomy and distribution of the vascular plants of the British Isles.
Address: Zoological Gardens, Regent's Park, London, NW1 4RY

17 **Victorian Society** *(1958 – see under 1)*

18 **Noise Abatement Society**
Founded 1959. The Society's purpose is to eliminate all excessive and unnecessary noise from all sources by taking all possible steps under the existing law to protect the public from assault by noise; to inform the public by every means of the dangers of noise to health and pocket and of their legal rights against those who create noise; to press for enforcement of present laws against noise and for new by-laws where existing laws appear inadequate.
Address: 6 Old Bond Street, London, W1X 3TA

19 **Conservation Society**
Founded 1966. Its objects are to secure for humanity a way of life which shall satisfy human needs and aspirations and be capable of indefinite continuance. It is especially concerned with problems of population size, conservation of animal, vegetable, and mineral resources, the protection of the environment and the prevention of pollution.
Address: 12 London Street, Chertsey, Surrey, KT16 8AA

OBSERVERS

1 **Royal Society for the Encouragement of Arts, Manufactures and Commerce (RSA)**
Founded in 1754, the Society has been closely associated with environmental problems throughout its history. The Society's Committee for the Environment reviews major developments in the environmental sphere at home and abroad, organizing meetings and conferences and taking such other action as may be appropriate. The Chairman of CoEnCo is normally a member of that Committee. The Society is particularly well placed to bring together for discussion the many sectors of society involved in environmental issues.
Address: John Adam Street, Adelphi, London, WC2N 6EZ

2 **Commons, Open Spaces and Footpaths Preservation Society**
Founded in 1865 as the Commons Society to save the London commons from enclosure. It soon extended its work to cover commons in other parts of the country and to public footpaths. Its main objects are to protect commons and greens for public use; to preserve existing footpaths and bridleways, secure the creation of new public rights of way and obtain public access to open country.
Address: Suite 4, 166 Shaftesbury Avenue, London, WC2H 8JH

3 **Town and Country Planning Association**
Founded 1899. By means of reports, conferences, and a monthly journal it advocates, and promotes an understanding of, national and regional planning policies that will improve living and working conditions, safeguard the best countryside and farmland, enhance natural, architectural and cultural amenities, and advance economic efficiency.
Address: 17 Carlton House Terrace, London, SW1Y 5AS

4 **The Pedestrians' Association for Road Safety**
Founded 1929. To ensure the safety of the public generally on the highways and elsewhere, and the freedom, comfort and general amenities of pedestrians. The Association has also been traditionally concerned with wider issues of transport planning, standing for sanity, safety and the protection of the environment. Public meetings are held in the Spring and Autumn. The Association's journal *Arrive* is published three times a year.
Address: 166 Shaftesbury Avenue, London, WC2H 8JH

5 **Nature Conservancy Council**
The Nature Conservancy Council was established by Act of Parliament in 1973 as an independent body under the Department of the Environment. Its functions are to establish, maintain and manage nature reserves in Great Britain; to provide advice to Ministers on the development and implementation of policies for or affecting nature conservation in Great Britain; to provide advice and disseminate knowledge about nature conservation, which means the conservation of flora, fauna or geological or physiographical features; and to commission or support research relevant to the above activities. (The Council replaces the Nature Conservancy, which was created in 1949 and which, since 1965, was a component body of the Natural Environment Research Council.)
Address: 19-20 Belgrave Square, London, SW1X 8PY

DESIGNED BY CECIL F. MEYER

PRINTED IN ENGLAND BY THE FURNIVAL PRESS · LONDON